Summary of

Tribe

On Homecoming and Belonging

by Sebastian Junger

Instaread

Instaread on Tribe

Please Note

This is a summary with analysis.

Copyright © 2016 by Instaread. All rights reserved worldwide. No part of this publication may be reproduced or transmitted in any form without the prior written consent of the publisher.

Limit of Liability/Disclaimer of Warranty: The publisher and author make no representations or warranties with respect to the accuracy or completeness of these contents and disclaim all warranties such as warranties of fitness for a particular purpose. The author or publisher is not liable for any damages whatsoever. The fact that an individual or organization is referred to in this document as a citation or source of information does not imply that the author or publisher endorses the information that the individual or organization provided. This concise summary is unofficial and is not authorized, approved, licensed, or endorsed by the original book's author or publisher.

Table of Contents

Overview 4

Important People 6

Key Takeaways 7

Analysis 10

Key Takeaway 1 10

Key Takeaway 2 12

Key Takeaway 3 14

Key Takeaway 4 15

Key Takeaway 5 17

Key Takeaway 6 19

Key Takeaway 7 21

Key Takeaway 8 23

Author's Style 25

Author's Perspective 27

References 29

Overview

Tribe by Sebastian Junger is a scientific and journalistic consideration of the correlation between societies with egalitarian tribal structures and low rates of mental illness, particularly post-traumatic stress disorder (PTSD) in soldiers returning home.

The sense of tribal belonging was documented in the eighteenth century among settlers in North America, who often joined Native American tribes even after those tribes held them as prisoners or waged war against the settlers. Those tribes were particularly egalitarian in nature, and despite lacking what were then modern amenities, members seldom worked as hard as the settlers in towns. The tribes also had low rates of depression and suicide.

Crisis in a community, whether that crisis is war or a natural disaster, tends to inspire people to return to a more collaborative, tribal mentality by sharing their resources regardless of social divisions and by working to help each other. During these crises, mental health markers also tend to improve. Men are more likely to risk their lives for others, and women are more likely to receive the benefits of life-risking heroics and to offer assistance when they empathize with victims in a crisis.

PTSD is a natural response to stress, especially when soldiers witness harm visited on others even if they were never in danger themselves. Serious symptoms are likely to occur when soldiers struggle to transition back to life among civilians. Israelis have remarkably low rates of PTSD, possibly because such a large proportion of the population has been in the military and therefore understand their fellow veterans' experience. On the other hand, anything that encourages former combatants to hold onto a sense of victimization causes them to recover from trauma more slowly.

Tribal societies ease the transition from war to civilian life through egalitarian, cohesive communities that make ex-combatants feel useful and do not assign them victim status. In societies where communities are divided and selfish people can steal from the community with relative impunity, that transition is more difficult and can result in violence by community members. Veterans particularly need ways to communicate their emotions to the civilian community and to experience unity among the people on whose behalf they fought.

Important People

Sebastian Junger is a best-selling author and documentary filmmaker known for reporting on war and natural disasters.

Rachel Yehuda directs the Traumatic Stress Studies Division at the Mount Sinai School of Medicine in New York.

Nidžara Ahmetašević is a Bosnian journalist who lived through the Bosnian War in Sarajevo.

Gregory Gomez is a member of the Apache tribe who fought in the Vietnam War and struggled with PTSD when he returned to the United States.

Key Takeaways

1. Tribal culture appeals to human beings on a deeply ingrained level. People will often choose to live with fewer modern amenities in exchange for the equality, shared labor, and companionship.

2. War and crisis can temporarily trigger an increased sense of community and generosity accompanied by decreases in suicide rates and rates of depression.

3. Men and women differ in the ways they respond to crises with men more often committing heroics to save women and children and women being moved by empathy to meet others' emotional needs.

4. Symptoms of PTSD arise in people who experience trauma regardless of whether they were involved in combat or were ever in danger. PTSD is most apparent when soldiers transition from war to civilian life.

5. PTSD recovery is hindered when soldiers feel alienated by their experiences, can identify no purpose for their sacrifices, or feel useless when they return home.

6. Societies with relatively low rates of PTSD transition soldiers back into society more carefully,

offer a cohesive and egalitarian environment, do not identify them as victims, and give them purpose.

7. In the United States, where a small proportion of the population serves in the military, veterans can feel isolated because so few people understand their experiences. Divisive rhetoric and unpunished greed prevent them from identifying with the people for whom they fought.

8. Veterans most of all need ways to share their emotional burdens with civilians, should feel like part of a united community, and should see that people who take advantage of the system are held accountable.

Thank you for purchasing this Instaread book

Download the Instaread mobile app to get unlimited text & audio summaries of bestselling books.

Visit Instaread.co to learn more.

Analysis

Key Takeaway 1

Tribal culture appeals to human beings on a deeply ingrained level. People will often choose to live with fewer modern amenities in exchange for the equality, shared labor, and companionship.

Analysis

When Europeans settled in towns in North America, they had regular encounters with indigenous tribes and often fought with those tribes over resources. Despite this contentious relationship, Europeans regularly left their towns to live with the tribes and even joined the tribe after natives captured them. Few natives ever left a tribe to join colonial society, however. Early accounts of indigenous tribal life seldom mentioned suicide and even less often mentioned depression-related suicide.

Tribal society in the United States is very different today than it was in the sixteenth and seventeenth centuries. It more closely resembles the rest of American society in its economic pressures with the added disadvantage of the uneven overlap of criminal justice and health services between the US government and tribal governments. Poverty, suicide, and diseases resulting from alcohol abuse are disproportionately common in the Native American population, as is diabetes. [1] Given the displacement and the extreme violence that Native American communities experienced as colonists expanded across the continent, it is not surprising that Native American tribes lost many of the quality of life advantages of their former tribal cultures. Native Americans were forced into non-tribal education, their children were adopted into white families, and their tribal religious practices were sometimes prohibited. Quality of life in Native American communities suffered as a result.

Key Takeaway 2

War and crisis can temporarily trigger an increased sense of community and generosity accompanied by decreases in suicide rates and rates of depression.

Analysis

During World War II, people renewed community ties and were more willing to share resources. Despite the stress of bombing in London, data shows that evidence of psychiatric distress, particularly depression and suicide, dropped until the end of the war. This trend is common in times of war. The growth of tribal feelings within the community could be the cause of the public welfare initiatives that became law immediately after the end of the war.

Over time, the feelings of community have decreased in many regions as an emphasis on authority has increased, a fact that is reflected by the tendency of authoritarian leaders to manipulate tribal and community dynamics in order to control a population. Authoritarian leaders often intentionally form rifts in society, an activity that constitutes the ultimate betrayal of the tribe and a crime that would be worthy of harsh punishment. Both Adolf Hitler and Josef Stalin designated large portions of their countries' populations as unfit for society, most notably Jews, political dissidents, homosexuals, and the European Roma, also called gypsies. As president of Iraq, Saddam

Hussein waged persecution campaigns against Shia Muslims, the Kurdish people, and the Marsh Arab tribes, all to quash dissent and increase his own power. [2] With the evidence supporting the mental health benefits of greater community involvement, it is worthwhile to consider the advantages of a cohesive, egalitarian tribe when others engage in rhetoric that attempts to vilify a subset of the community and unite groups against immigrants, religions, or political parties.

Key Takeaway 3

Men and women differ in the ways they respond to crises with men more often committing heroics to save women and children and women being moved by empathy to meet others' emotional needs

Analysis

Gender roles influence differences in the ways that men and women exhibit heroism. When others are not present to fulfill a particular role, men and women can adapt their gender roles; women may commit acts of heroism and men may assist the needy. Gender roles are thus influenced but not determined by sex.

Historical examples of ruling couples and women who succeed men as leaders in war reinforce the notion that members of both sexes adapt in response to crisis. When Ethiopia fought against attempted Italian colonization in 1895, Emperor Menelik managed challenging diplomatic situations and treaties while his wife, Empress Taytu Betul, became a cunning strategist, a confrontational negotiator, and a respected military leader. [3] Remedios Gomez-Paraiso became the commander of resistance fighters in the Philippines during World War II after Japanese soldiers publicly killed her father, a resistance organizer. [4] Although gender roles have historically been enforced by social mores, sex difference cannot reliably be used to predict how anyone will act when a community is in danger.

Key Takeaway 4

Symptoms of PTSD arise in people who experience trauma regardless of whether they were involved in combat or were ever in danger. PTSD is most apparent when soldiers transition from war to civilian life.

Analysis

People who conduct drone strikes remotely and other service members who do not directly experience combat are just as likely to experience PTSD as people who experience combat firsthand because they all see the harm that war does even if they were never in danger themselves. PTSD symptoms are less common among people who are still deployed but more common when they end their tours of duty and return to the United States.

On top of the trauma that results in PTSD among non-combat veterans, treatment for PTSD can cause traumatization to the people involved in veterans' treatment. This is called secondary trauma, vicarious traumatization, or secondary traumatization, and it refers to the trauma non-combat treatment staff experience when they see and discuss the trauma that soldiers witnessed firsthand. This phenomenon is also called compassion fatigue, and it is particularly common among psychiatrists, counselors, therapists, and medical personnel. [5] This vicarious trauma was even blamed for a rampage shooting at Fort Hood, Texas, in 2009, when an Army psychiatrist killed 13 people and wounded 30. The suspect never was deployed or saw combat, but he worked daily with soldiers who were struggling to recover from PTSD. [6] Given that the transition from traumatic wartime experience to civilian life is an important factor in managing PTSD symptoms, psychiatrists would be especially vulnerable because they transition between the stressful environment of treating soldiers back to the civilian world every day. They likely have little time to cope with those emotions aside from a commute and little access to specialists who understand and can treat compassion fatigue.

Key Takeaway 5

PTSD recovery is hindered when soldiers feel alienated by their experiences, can identify no purpose for their sacrifices, or feel useless when they return home.

Analysis

The societies with the lowest rates of PTSD are the ones where everyone, even civilians, understands the trauma of military service. Low-PTSD societies can also be those in which civilians also feel a constant sense of danger and feel that their lives are at risk. Above all, societies that ensure veterans will be employed thereby help to prevent PTSD because employed veterans feel they are needed and useful rather than a burden on the community.

Veteran unemployment affects other matters of US veteran health and wellness, particularly their housing and mental health treatment. As of 2015, the unemployment rate for veterans of active duty since September 2001 was 5.8 percent, compared to 5 percent for the general population. Veterans of all wars were unemployed at a rate of 4.6 percent, slightly less than average, but the rate for female veterans alone was 5.4 percent. [7] Homeless people in the United States are disproportionately likely to be veterans and also disproportionately likely to be minority veterans. The Department of Housing and Urban Development estimates that, at any given time, about 47,725 veterans are homeless, and an additional 1.4 million are at risk of homelessness. While they are homeless, veterans have inconsistent access to treatment for PTSD and substance abuse, making them less employable. [8] To address this, the US government has established programs to encourage employers to hire veterans and numerous programs to support wounded veterans. There are also innovative programs such as Mission Continues, founded by a veteran Marine, which connects wounded post-September 2001 veterans with volunteer opportunities that give them a sense of purpose despite being unable to serve in the military anymore. [9]

Key Takeaway 6

Societies with relatively low rates of PTSD transition soldiers back into society more carefully, offer a cohesive and egalitarian environment, do not identify them as victims, and give them purpose.

Analysis

Rituals practiced in indigenous tribes encouraged a gradual, community-involved transition back to normal life for returning soldiers and offered spiritual programs of treatment that were also community-involved. Iroquois soldiers were expected to contribute to the community when they returned as well.

Israel may have unusually low rates of PTSD connected in part to the constant existential threat that civilians and veterans feel in everyday society and to the fact that military service is mandatory for most Israeli citizens. However, there are significant numbers of minorities whose experiences as Israeli Defense Force soldiers differ from that of Israeli Jews. Members of Arab populations, such as the Druze, Bedouins, and Palestinian Christians, regularly sign up for IDF service. The Druze are included in the mandatory conscription laws and Bedouins are encouraged to volunteer for the economic advantages. Muslim Arabs of Israel and Palestine also volunteer for the IDF, at lower numbers. As soldiers, Arabs in the IDF often experience unequal treatment from their fellow soldiers. When they return home after their service, the Israelis with whom they served do not usually maintain their relationship, and the Arab soldiers in their own communities lack the support networks and shared experience of service that help Israeli Jews recover from PTSD after their service [10]. This disparity suggests that it's likely that Arab IDF soldiers experience a higher rate of PTSD than the general population of IDF veterans.

Key Takeaway 7

In the United States, where a small proportion of the population serves in the military, veterans can feel isolated because so few people understand their experiences. Divisive rhetoric and unpunished greed prevent them from identifying with the people for whom they fought.

Analysis

In the United States, just 1 percent of citizens serve in the military, meaning that soldiers who return home have few people who understand what they experienced and who can support them as they recover. Similarly, soldiers who return from war may struggle to understand whether the results of their service are worth their sacrifice. For example, individuals who take advantage of the community for personal gains, such as those who commit large-scale theft and fraud, give the impression that they receive a disproportionately large amount of the national security benefit from the armed forces, and they often then misuse that benefit to protect their ill-gotten gains. Moreover, divisive political rhetoric may unite people against a common enemy, which research suggests is one cause of the findings on shifts to tribal communities during war against a common enemy, but in peacetime it also forms rifts domestically between citizens who would be best served cooperating together toward a common goal.

The dissonance between the sacrifices that US culture expects from soldiers and the support it gives when they come home is more frequently addressed in literary fiction. Ben Fountain's 2012 novel *Billy Lynn's Long Halftime Walk* follows a squad of soldiers returning to the United States from Iraq to be honored for their bravery. Along the way, the soldiers encounter characters who reflect the sometimes hypocritical or inappropriately focused attitudes of civilians when they discuss war. The book won the National Book Critics Circle Award for Fiction in 2012. [11] In 2014, Phil Klay won the National Book Award in Fiction for *Redeployment*, a short story collection that presented a variety of fictional soldiers' experiences, based on Klay's service and that of veterans he interviewed. The book features stories of both soldiers overseas, in the heart of warrior culture, and of soldiers attempting to transition back to their lives in the civilian world, where they sometimes struggle to communicate what they feel or what they experienced to people who never served. [12]

Key Takeaway 8

Veterans most of all need ways to share their emotional burdens with civilians, should feel like part of a united community, and should see that people who take advantage of the system are held accountable.

Analysis

In addition to more obvious actions, such as ensuring better veteran employment and health care, creating a better support system in the United States would involve some fundamental changes in the way civilians communicate with soldiers, particularly the way they listen. Some events that foster open communication between soldiers and civilians have already occurred in a town hall format and these could become more commonplace. Civilians should mitigate divisive rhetoric when it damages the sense of community that inspires generosity and equality toward others including veterans. Additionally, people who take advantage of the system, such as white-collar embezzlers who steal millions and sometimes billions of dollars, would need to be more fairly punished for the damage they do to communities.

One place where these changes would have the most impact is Guam. On the island, one in eight adults has served in the US military, but as a US territory, Guam has just one non-voting delegate in Congress and has no electoral college votes in the presidential election. In 2014,

Guam received the lowest amount of medical spending from the Department of Veterans Affairs per capita of all states and territories, and it was not well-staffed to serve veterans with PTSD. Until recent changes in the VA system, Guam veterans were required to travel 3,000 miles to Hawaii to access an intensive treatment program for PTSD. [13] While they could find comfort in a shared experience, given the higher than average rate of military service, Guamanians would not feel equal to the mainland US residents whose votes actually affect US policy, so the sacrifice they made has less meaning. Similarly, the lack of accessible PTSD treatment programs is an obstacle to healthy recovery because PTSD treatment can address underlying causes of symptoms, such as substance abuse or childhood trauma, that tribal community structures are less equipped to handle.

Author's Style

Sebastian Junger writes in a journalistic style throughout *Tribe*. The text is strongly supported by statistics from scientific studies. It also contains historical accounts, quotes from sources Junger interviewed, and personal anecdotes from Junger's life, mostly from his work as a war correspondent in Bosnia and Afghanistan. These personal accounts are highly detailed and often morbid due to the subject matter. Junger describes scenes of war and natural disasters that he witnessed and from history in vivid detail.

Junger writes in a concise, direct style that moves quickly without much transition between topics. The diction is approachable and Junger uses a somber tone appropriate for the subject matter. There is no technical language despite large portions of the book spent recounting military encounters and psychiatric concepts.

Junger focuses on history and phenomena in Native American, Israeli, and eastern European communities. He offers balanced perspective laying out the evidence for and against the connections he suggests and mentioning the factors that complicate findings in each case. He does not offer advice for the reader, but some of the findings could be interpreted as policy suggestions.

Junger acknowledges early on that he chose a controversial term for indigenous American tribes, American Indians, because one of his sources insisted that he use the phrase, even though it can be confusing or regarded as politically incorrect. When describing these indigenous tribes' cultures, Junger uses and defines words from their native languages and discusses them with respect, not exoticism.

Author's Perspective

Sebastian Junger began his journalism career reporting on the Bosnian War from Sarajevo in the early 1990s, an experience that he mentions frequently in *Tribe*. Junger also worked as a war correspondent in Afghanistan, which resulted in his books *Fire* (2002) and *War* (2010) and his documentary films *Restrepo* (2009), *Which Way Is the Front Line From Here?* (2013), *Korengal* (2014), and *The Last Patrol* (2014). Junger also wrote *The Perfect Storm: A True Story of Men Against the Sea* (1997), a widely acclaimed creative non-fiction book about the events that resulted in the disappearance of a fishing boat near Nova Scotia during a storm in 1991. Junger is also well-known for his work alongside photographer Tim Hetherington, who died while covering the Libyan civil war in 2011, and is often involved in projects to commemorate him. Hetherington does not appear in *Tribe* but would no doubt have accompanied Junger for the events in Afghanistan that he recounts.

~~~~ **END OF INSTAREAD** ~~~~

Instaread on Tribe

Thank you for purchasing this Instaread book

Download the Instaread mobile app to get unlimited text & audio summaries of bestselling books.

Visit Instaread.co to learn more.

References

1. Gordon, Claire. "5 big Native American health issues you don't know about." *Al Jazeera America Flagship Blog*. August 28, 2013. Updated May 29, 2014. Accessed June 8, 2016. http://america.aljazeera.com/watch/shows/america-tonight/america-tonight-blog/2013/8/28/5-huge-native-americanhealthissuesyoudontknowabout.html

2. Johns, Dave. "The Crimes of Saddam Hussein." *Frontline World*. January 24, 2006. Accessed June 25, 2016. http://www.pbs.org/frontlineworld/stories/iraq501/events_index.html

3. Pankhurst, Rita. "Women of Power in Ethiopia: Elegance and Power." *One World*. 1995. Accessed June 26, 2016. http://www.oneworldmagazine.org/focus/etiopia/women3.html

4. Orejas, Tonette. "Liwayway: Warrior who wore lipstick in gun battles." *Inquirer.net*. May 17, 2014. Accessed June 26, 2016. http://newsinfo.inquirer.net/602758/liwayway-warrior-who-wore-lipstick-in-gun-battles

5. Simpson, Laura, and Donna Starkey. "Secondary Traumatic Stress, Compassion Fatigue, and Counselor Spirituality: Implications for Counselors Working in Trauma." Counseling.org. 2006. Accessed June 8, 2016. https://www.

counseling.org/resources/library/Selected%20Topics/Crisis/Simpson.htm

6. Landau, Elizabeth. "Treating trauma victims may cause its own trauma." *CNN*. November 7, 2009. Accessed June 8, 2016. http://www.cnn.com/2009/HEALTH/11/06/military.psychiatrists.fort.hood/index.html?iref=24hours

7. Press Office. "Employment Situation of Veterans Summary." Bureau of Labor Statistics. March 22, 2016. Accessed June 8, 2016. http://www.bls.gov/news.release/vet.nr0.htm

8. "Background & Statistics." National Coalition for Homeless Veterans. Accessed June 8, 2016. http://nchv.org/index.php/news/media/background_and_statistics/

9. "Our Story." The Mission Continues. Accessed June 8, 2016. https://www.missioncontinues.org/about/history/

10. Hoffman, Jordan. "His deep, dark secret: He's Arab, Muslim, and serves in the IDF." *The Times of Israel*. November 10, 2012. Accessed June 8, 2016. http://www.timesofisrael.com/his-big-secret-hes-arab-muslim-and-serves-in-the-idf/

11. Chiusano, Mark. "Amid NFL Scandals, A Novel About America's Enduring Love Of The Sport." *NPR Books*. September 19, 2014.

Accessed June 8, 2016. http://www.npr.org/2014/09/19/349865898/amidst-nfl-scandals-a-novel-about-america-s-enduring-love-of-the-sport

12. "2014 National Book Award Winner, Fiction: Phil Klay." National Book Foundation. Accessed June 8, 2016. http://www.nationalbook.org/nba2014_f_klay.html

13. "US Territories." *Last Week Tonight*. HBO. March 8, 2015. Accessed June 8, 2016. https://www.youtube.com/watch?v=CesHr99ezWE

Lightning Source UK Ltd.
Milton Keynes UK
UKOW06f1808250617
304071UK00018B/466/P